**STRAIGHT
FORWARD
MATH**

Division

by S. Harold Collins

Book design and illustrations by
Kathy Kifer

Published by
Garlic Press
605 Powers Street
Eugene, OR 97402

ISBN: 978-0-931993-13-8
Order Number GP-013
Printed in China

www.garlicpress.com

To Parents and Teachers

The Straight Forward Math Series has been designed for parents and teachers of children. It is a simple, straightforward way to teach division facts 1 – 10.

The Straight Forward Math Series gives practice, review, and testing. Use these steps to teach basic division skills.

- Give the **Beginning Assessment Test** to find out where to begin with Practice Sheets. The Beginning Assessment Test (page 7) will tell you which facts are sound and which need attention.

  Look closely at the Beginning Assessment Test. Division facts are diagonally arranged (see Answers, page 34 for the display of facts).

  If the Beginning Assessment Test is given and errors begin with 2s, that is where you begin with Practice Sheets to build skills.

- Start **Practice Sheets** at the appropriate error level. Do not skip levels once you begin. Build up to 10s facts.

  Two Practice Sheets are given for each level to provide ample practice. Each Practice Sheet has 100 problems.

  Set a standard to move to the next division level; for instance, 95 problems correctly answered out of 100 in four minutes.

  Two forms are used for Practice Sheets— $3)\overline{21}$ and $21 \div 3 = $. Both forms occur commonly.

- Give **Review Sheets** after completion of facts through 4, facts through 7, and again after facts through 10.

- Give a **Section Diagnostic Test** as a final measure for a particular section. Section Diagnostic Tests are arranged to identify problems that may still exist with particular division facts (much like the original Beginning Assessment Test).

  Set a standard to move from one section to the next. If the standard is not met, go back, focus on problem area(s) with Practice Sheets or similar material you can prepare.

- Give the **Final Assessment Test** to measure all facts 1 – 10. Compare the change from the Beginning Assessment Test.

Contents

Beginning Assessment Test

3)3 2)4 1)3 4)8 3)0 2)8 1)2 8)16 9)18 10)10

4)4 3)6 2)2 1)2 4)4 3)3 2)12 1)10 8)56 9)27

5)0 4)12 3)9 2)0 1)4 4)12 3)6 2)2 1)0 8)64

6)12 5)5 4)8 3)12 2)10 1)5 4)20 3)9 2)4 1)3

7)7 6)6 5)10 4)0 3)15 2)14 1)6 4)40 3)18 2)6

8)16 7)14 6)18 5)15 4)16 3)3 2)16 1)7 4)16 3)30

9)81 8)8 7)21 6)48 5)20 4)20 3)21 2)18 1)8 4)24

2)12 9)36 8)24 7)28 6)24 5)25 4)24 3)24 2)20 1)9

10)30 2)8 9)45 8)32 7)49 6)36 5)30 4)28 3)27 2)12

10)20 10)40 2)18 9)54 8)40 7)42 6)42 5)35 4)32 3)18

Dividing 1 and 2

$1\overline{)1}$ $2\overline{)2}$ $1\overline{)2}$ $2\overline{)4}$ $1\overline{)3}$ $2\overline{)6}$ $1\overline{)4}$ $2\overline{)8}$ $1\overline{)5}$ $2\overline{)10}$

$1\overline{)6}$ $2\overline{)12}$ $1\overline{)7}$ $2\overline{)14}$ $1\overline{)8}$ $2\overline{)0}$ $1\overline{)9}$ $2\overline{)18}$ $1\overline{)10}$ $2\overline{)20}$

$2\overline{)10}$ $1\overline{)5}$ $2\overline{)8}$ $1\overline{)4}$ $1\overline{)3}$ $2\overline{)6}$ $2\overline{)4}$ $2\overline{)2}$ $1\overline{)2}$ $2\overline{)14}$

$1\overline{)9}$ $1\overline{)1}$ $2\overline{)12}$ $2\overline{)16}$ $1\overline{)6}$ $1\overline{)7}$ $2\overline{)20}$ $1\overline{)10}$ $2\overline{)18}$ $1\overline{)0}$

$2\overline{)20}$ $2\overline{)16}$ $1\overline{)3}$ $2\overline{)8}$ $1\overline{)5}$ $2\overline{)10}$ $1\overline{)0}$ $2\overline{)6}$ $1\overline{)7}$ $2\overline{)14}$

$1\overline{)2}$ $2\overline{)12}$ $1\overline{)0}$ $1\overline{)1}$ $1\overline{)10}$ $2\overline{)10}$ $1\overline{)4}$ $2\overline{)4}$ $1\overline{)6}$ $1\overline{)8}$

$2\overline{)20}$ $1\overline{)7}$ $2\overline{)18}$ $1\overline{)8}$ $2\overline{)16}$ $1\overline{)9}$ $2\overline{)14}$ $1\overline{)10}$ $2\overline{)12}$ $1\overline{)10}$

$1\overline{)10}$ $1\overline{)1}$ $1\overline{)8}$ $1\overline{)2}$ $2\overline{)6}$ $1\overline{)3}$ $2\overline{)4}$ $1\overline{)2}$ $2\overline{)2}$ $2\overline{)0}$

$2\overline{)4}$ $1\overline{)0}$ $2\overline{)6}$ $1\overline{)9}$ $2\overline{)8}$ $1\overline{)8}$ $2\overline{)10}$ $1\overline{)7}$ $2\overline{)12}$ $1\overline{)6}$

$2\overline{)14}$ $1\overline{)5}$ $2\overline{)16}$ $1\overline{)4}$ $2\overline{)18}$ $1\overline{)3}$ $2\overline{)20}$ $1\overline{)2}$ $2\overline{)0}$ $1\overline{)1}$

Dividing 1 and 2

1 ÷ 1 =	2 ÷ 1 =	3 ÷ 1 =	4 ÷ 1 =	5 ÷ 1 =
10 ÷ 2 =	8 ÷ 2 =	6 ÷ 2 =	4 ÷ 2 =	2 ÷ 1 =
6 ÷ 1 =	7 ÷ 1 =	8 ÷ 1 =	9 ÷ 1 =	10 ÷ 1 =
9 ÷ 1 =	12 ÷ 2 =	6 ÷ 1 =	20 ÷ 2 =	18 ÷ 2 =
20 ÷ 2 =	3 ÷ 1 =	5 ÷ 1 =	9 ÷ 1 =	7 ÷ 1 =
2 ÷ 1 =	0 ÷ 1 =	10 ÷ 1 =	4 ÷ 1 =	6 ÷ 1 =
14 ÷ 2 =	18 ÷ 2 =	16 ÷ 2 =	14 ÷ 2 =	12 ÷ 2 =
10 ÷ 1 =	8 ÷ 1 =	6 ÷ 2 =	4 ÷ 2 =	2 ÷ 2 =
4 ÷ 2 =	6 ÷ 2 =	8 ÷ 2 =	2 ÷ 1 =	12 ÷ 2 =
14 ÷ 2 =	16 ÷ 2 =	18 ÷ 2 =	20 ÷ 2 =	0 ÷ 2 =
5 ÷ 1 =	4 ÷ 2 =	3 ÷ 1 =	8 ÷ 2 =	10 ÷ 2 =
12 ÷ 2 =	14 ÷ 2 =	0 ÷ 2 =	18 ÷ 2 =	20 ÷ 2 =
5 ÷ 1 =	4 ÷ 1 =	6 ÷ 1 =	2 ÷ 1 =	1 ÷ 1 =
8 ÷ 1 =	16 ÷ 2 =	7 ÷ 1 =	10 ÷ 2 =	0 ÷ 1 =
16 ÷ 2 =	8 ÷ 2 =	10 ÷ 2 =	6 ÷ 2 =	14 ÷ 2 =
12 ÷ 2 =	1 ÷ 1 =	3 ÷ 1 =	4 ÷ 2 =	8 ÷ 1 =
7 ÷ 1 =	8 ÷ 1 =	9 ÷ 1 =	10 ÷ 1 =	5 ÷ 1 =
6 ÷ 2 =	2 ÷ 1 =	10 ÷ 2 =	2 ÷ 2 =	0 ÷ 2 =
0 ÷ 1 =	9 ÷ 1 =	8 ÷ 1 =	7 ÷ 1 =	6 ÷ 1 =
2 ÷ 2 =	12 ÷ 2 =	0 ÷ 1 =	10 ÷ 2 =	14 ÷ 2 =

Dividing 3

$3\overline{)3}$ $3\overline{)6}$ $3\overline{)0}$ $3\overline{)9}$ $3\overline{)12}$ $3\overline{)18}$ $3\overline{)15}$ $3\overline{)24}$ $3\overline{)27}$ $3\overline{)21}$

$3\overline{)18}$ $3\overline{)24}$ $3\overline{)6}$ $3\overline{)27}$ $3\overline{)3}$ $3\overline{)12}$ $3\overline{)15}$ $3\overline{)9}$ $3\overline{)21}$ $3\overline{)30}$

$3\overline{)15}$ $3\overline{)3}$ $3\overline{)24}$ $3\overline{)30}$ $3\overline{)12}$ $3\overline{)6}$ $3\overline{)18}$ $3\overline{)27}$ $3\overline{)21}$ $3\overline{)9}$

$3\overline{)6}$ $3\overline{)15}$ $3\overline{)12}$ $3\overline{)21}$ $3\overline{)3}$ $3\overline{)30}$ $3\overline{)18}$ $3\overline{)24}$ $3\overline{)9}$ $3\overline{)0}$

$3\overline{)27}$ $3\overline{)15}$ $3\overline{)3}$ $3\overline{)12}$ $3\overline{)21}$ $3\overline{)9}$ $3\overline{)24}$ $3\overline{)30}$ $3\overline{)6}$ $3\overline{)27}$

$3\overline{)9}$ $3\overline{)12}$ $3\overline{)21}$ $3\overline{)3}$ $3\overline{)24}$ $3\overline{)30}$ $3\overline{)6}$ $3\overline{)9}$ $3\overline{)27}$ $3\overline{)15}$

$3\overline{)18}$ $3\overline{)21}$ $3\overline{)6}$ $3\overline{)30}$ $3\overline{)15}$ $3\overline{)24}$ $3\overline{)3}$ $3\overline{)27}$ $3\overline{)12}$ $3\overline{)9}$

$3\overline{)6}$ $3\overline{)30}$ $3\overline{)15}$ $3\overline{)21}$ $3\overline{)3}$ $3\overline{)18}$ $3\overline{)24}$ $3\overline{)9}$ $3\overline{)27}$ $3\overline{)12}$

$3\overline{)12}$ $3\overline{)24}$ $3\overline{)9}$ $3\overline{)18}$ $3\overline{)30}$ $3\overline{)21}$ $3\overline{)3}$ $3\overline{)27}$ $3\overline{)15}$ $3\overline{)6}$

$3\overline{)0}$ $3\overline{)6}$ $3\overline{)15}$ $3\overline{)27}$ $3\overline{)12}$ $3\overline{)24}$ $3\overline{)30}$ $3\overline{)3}$ $3\overline{)9}$ $3\overline{)18}$

Dividing 3

3 ÷ 3 =	0 ÷ 3 =	12 ÷ 3 =	15 ÷ 3 =	27 ÷ 3 =
18 ÷ 3 =	6 ÷ 3 =	3 ÷ 3 =	18 ÷ 3 =	21 ÷ 3 =
15 ÷ 3 =	24 ÷ 3 =	12 ÷ 3 =	24 ÷ 3 =	9 ÷ 3 =
6 ÷ 3 =	12 ÷ 3 =	3 ÷ 3 =	15 ÷ 3 =	6 ÷ 3 =
27 ÷ 3 =	3 ÷ 3 =	21 ÷ 3 =	6 ÷ 3 =	21 ÷ 3 =
9 ÷ 3 =	21 ÷ 3 =	24 ÷ 3 =	18 ÷ 3 =	27 ÷ 3 =
18 ÷ 3 =	6 ÷ 3 =	15 ÷ 3 =	3 ÷ 3 =	12 ÷ 3 =
6 ÷ 3 =	15 ÷ 3 =	3 ÷ 3 =	24 ÷ 3 =	15 ÷ 3 =
12 ÷ 3 =	9 ÷ 3 =	30 ÷ 3 =	3 ÷ 3 =	9 ÷ 3 =
0 ÷ 3 =	15 ÷ 3 =	12 ÷ 3 =	30 ÷ 3 =	27 ÷ 3 =
6 ÷ 3 =	27 ÷ 3 =	24 ÷ 3 =	24 ÷ 3 =	18 ÷ 3 =
24 ÷ 3 =	18 ÷ 3 =	21 ÷ 3 =	9 ÷ 3 =	6 ÷ 3 =
3 ÷ 3 =	21 ÷ 3 =	18 ÷ 3 =	27 ÷ 3 =	12 ÷ 3 =
15 ÷ 3 =	30 ÷ 3 =	24 ÷ 3 =	24 ÷ 3 =	9 ÷ 3 =
12 ÷ 3 =	3 ÷ 3 =	30 ÷ 3 =	30 ÷ 3 =	15 ÷ 3 =
15 ÷ 3 =	12 ÷ 3 =	9 ÷ 3 =	9 ÷ 3 =	3 ÷ 3 =
21 ÷ 3 =	21 ÷ 3 =	0 ÷ 3 =	27 ÷ 3 =	0 ÷ 3 =
30 ÷ 3 =	30 ÷ 3 =	6 ÷ 3 =	9 ÷ 3 =	9 ÷ 3 =
24 ÷ 3 =	27 ÷ 3 =	12 ÷ 3 =	27 ÷ 3 =	30 ÷ 3 =
6 ÷ 3 =	9 ÷ 3 =	18 ÷ 3 =	3 ÷ 3 =	24 ÷ 3 =

Dividing 4

4s

$4\overline{)4}$ $\quad$ $4\overline{)12}$ $\quad$ $4\overline{)24}$ $\quad$ $4\overline{)20}$ $\quad$ $4\overline{)40}$ $\quad$ $4\overline{)28}$ $\quad$ $4\overline{)16}$ $\quad$ $4\overline{)32}$ $\quad$ $4\overline{)36}$ $\quad$ $4\overline{)8}$

$4\overline{)16}$ $\quad$ $4\overline{)4}$ $\quad$ $4\overline{)32}$ $\quad$ $4\overline{)12}$ $\quad$ $4\overline{)24}$ $\quad$ $4\overline{)40}$ $\quad$ $4\overline{)8}$ $\quad$ $4\overline{)36}$ $\quad$ $4\overline{)28}$ $\quad$ $4\overline{)20}$

$4\overline{)28}$ $\quad$ $4\overline{)16}$ $\quad$ $4\overline{)8}$ $\quad$ $4\overline{)32}$ $\quad$ $4\overline{)20}$ $\quad$ $4\overline{)4}$ $\quad$ $4\overline{)36}$ $\quad$ $4\overline{)24}$ $\quad$ $4\overline{)40}$ $\quad$ $4\overline{)12}$

$4\overline{)12}$ $\quad$ $4\overline{)4}$ $\quad$ $4\overline{)8}$ $\quad$ $4\overline{)28}$ $\quad$ $4\overline{)16}$ $\quad$ $4\overline{)32}$ $\quad$ $4\overline{)20}$ $\quad$ $4\overline{)0}$ $\quad$ $4\overline{)12}$ $\quad$ $4\overline{)36}$

$4\overline{)32}$ $\quad$ $4\overline{)36}$ $\quad$ $4\overline{)40}$ $\quad$ $4\overline{)36}$ $\quad$ $4\overline{)32}$ $\quad$ $4\overline{)40}$ $\quad$ $4\overline{)36}$ $\quad$ $4\overline{)32}$ $\quad$ $4\overline{)40}$ $\quad$ $4\overline{)32}$

$4\overline{)12}$ $\quad$ $4\overline{)32}$ $\quad$ $4\overline{)16}$ $\quad$ $4\overline{)4}$ $\quad$ $4\overline{)24}$ $\quad$ $4\overline{)8}$ $\quad$ $4\overline{)16}$ $\quad$ $4\overline{)40}$ $\quad$ $4\overline{)36}$ $\quad$ $4\overline{)20}$

$4\overline{)8}$ $\quad$ $4\overline{)28}$ $\quad$ $4\overline{)20}$ $\quad$ $4\overline{)36}$ $\quad$ $4\overline{)4}$ $\quad$ $4\overline{)12}$ $\quad$ $4\overline{)40}$ $\quad$ $4\overline{)16}$ $\quad$ $4\overline{)24}$ $\quad$ $4\overline{)0}$

$4\overline{)40}$ $\quad$ $4\overline{)4}$ $\quad$ $4\overline{)36}$ $\quad$ $4\overline{)8}$ $\quad$ $4\overline{)32}$ $\quad$ $4\overline{)16}$ $\quad$ $4\overline{)24}$ $\quad$ $4\overline{)20}$ $\quad$ $4\overline{)16}$ $\quad$ $4\overline{)28}$

$4\overline{)36}$ $\quad$ $4\overline{)0}$ $\quad$ $4\overline{)12}$ $\quad$ $4\overline{)40}$ $\quad$ $4\overline{)16}$ $\quad$ $4\overline{)20}$ $\quad$ $4\overline{)12}$ $\quad$ $4\overline{)24}$ $\quad$ $4\overline{)8}$ $\quad$ $4\overline{)4}$

$4\overline{)20}$ $\quad$ $4\overline{)12}$ $\quad$ $4\overline{)40}$ $\quad$ $4\overline{)36}$ $\quad$ $4\overline{)4}$ $\quad$ $4\overline{)24}$ $\quad$ $4\overline{)16}$ $\quad$ $4\overline{)4}$ $\quad$ $4\overline{)32}$ $\quad$ $4\overline{)16}$

Dividing 4

4 ÷ 4 =	24 ÷ 4 =	40 ÷ 4 =	16 ÷ 4 =	36 ÷ 4 =
16 ÷ 4 =	32 ÷ 4 =	24 ÷ 4 =	8 ÷ 4 =	28 ÷ 4 =
28 ÷ 4 =	8 ÷ 4 =	20 ÷ 4 =	36 ÷ 4 =	40 ÷ 4 =
12 ÷ 4 =	8 ÷ 4 =	16 ÷ 4 =	20 ÷ 4 =	12 ÷ 4 =
32 ÷ 4 =	40 ÷ 4 =	32 ÷ 4 =	36 ÷ 4 =	40 ÷ 4 =
12 ÷ 4 =	16 ÷ 4 =	24 ÷ 4 =	16 ÷ 4 =	36 ÷ 4 =
8 ÷ 4 =	20 ÷ 4 =	4 ÷ 4 =	40 ÷ 4 =	24 ÷ 4 =
40 ÷ 4 =	36 ÷ 4 =	32 ÷ 4 =	24 ÷ 4 =	16 ÷ 4 =
36 ÷ 4 =	12 ÷ 4 =	16 ÷ 4 =	12 ÷ 4 =	8 ÷ 4 =
20 ÷ 4 =	40 ÷ 4 =	4 ÷ 4 =	16 ÷ 4 =	32 ÷ 4 =
12 ÷ 4 =	20 ÷ 4 =	28 ÷ 4 =	32 ÷ 4 =	8 ÷ 4 =
4 ÷ 4 =	12 ÷ 4 =	40 ÷ 4 =	36 ÷ 4 =	20 ÷ 4 =
16 ÷ 4 =	32 ÷ 4 =	4 ÷ 4 =	24 ÷ 4 =	12 ÷ 4 =
4 ÷ 4 =	28 ÷ 4 =	32 ÷ 4 =	0 ÷ 4 =	36 ÷ 4 =
36 ÷ 4 =	36 ÷ 4 =	4 ÷ 4 =	32 ÷ 4 =	32 ÷ 4 =
28 ÷ 4 =	4 ÷ 4 =	8 ÷ 4 =	4 ÷ 4 =	20 ÷ 4 =
4 ÷ 4 =	36 ÷ 4 =	16 ÷ 4 =	20 ÷ 4 =	0 ÷ 4 =
0 ÷ 4 =	8 ÷ 4 =	20 ÷ 4 =	24 ÷ 4 =	28 ÷ 4 =
12 ÷ 4 =	40 ÷ 4 =	24 ÷ 4 =	8 ÷ 4 =	4 ÷ 4 =
32 ÷ 4 =	24 ÷ 4 =	0 ÷ 4 =	4 ÷ 4 =	16 ÷ 4 =

Review Sheet 1 – 4

$1\overline{)10}$ $4\overline{)16}$ $2\overline{)12}$ $3\overline{)9}$ $4\overline{)40}$ $2\overline{)8}$ $3\overline{)21}$ $4\overline{)36}$ $1\overline{)1}$ $3\overline{)27}$

$1\overline{)8}$ $1\overline{)0}$ $2\overline{)6}$ $3\overline{)24}$ $4\overline{)24}$ $2\overline{)10}$ $3\overline{)12}$ $4\overline{)32}$ $1\overline{)2}$ $3\overline{)24}$

$1\overline{)3}$ $3\overline{)24}$ $2\overline{)20}$ $3\overline{)18}$ $4\overline{)16}$ $2\overline{)16}$ $3\overline{)6}$ $4\overline{)28}$ $1\overline{)3}$ $2\overline{)0}$

$1\overline{)6}$ $6\overline{)60}$ $2\overline{)8}$ $3\overline{)3}$ $4\overline{)12}$ $3\overline{)21}$ $3\overline{)30}$ $4\overline{)24}$ $1\overline{)9}$ $3\overline{)18}$

$1\overline{)2}$ $2\overline{)20}$ $2\overline{)4}$ $3\overline{)12}$ $4\overline{)4}$ $2\overline{)6}$ $3\overline{)24}$ $4\overline{)40}$ $1\overline{)4}$ $3\overline{)15}$

$2\overline{)16}$ $4\overline{)40}$ $2\overline{)18}$ $3\overline{)27}$ $4\overline{)32}$ $2\overline{)2}$ $3\overline{)9}$ $4\overline{)0}$ $1\overline{)7}$ $4\overline{)20}$

$2\overline{)14}$ $2\overline{)18}$ $2\overline{)2}$ $3\overline{)30}$ $4\overline{)28}$ $2\overline{)12}$ $3\overline{)15}$ $4\overline{)8}$ $1\overline{)8}$ $4\overline{)24}$

$2\overline{)18}$ $3\overline{)21}$ $2\overline{)16}$ $3\overline{)21}$ $4\overline{)20}$ $2\overline{)18}$ $3\overline{)0}$ $4\overline{)12}$ $1\overline{)6}$ $4\overline{)28}$

$2\overline{)12}$ $4\overline{)32}$ $2\overline{)10}$ $3\overline{)15}$ $4\overline{)8}$ $2\overline{)20}$ $3\overline{)18}$ $4\overline{)16}$ $1\overline{)8}$ $4\overline{)32}$

$2\overline{)10}$ $3\overline{)30}$ $2\overline{)14}$ $3\overline{)6}$ $4\overline{)24}$ $2\overline{)2}$ $3\overline{)27}$ $4\overline{)20}$ $1\overline{)10}$ $4\overline{)36}$

Section Diagnostic Test 1 – 4

20 ÷ 4 =	24 ÷ 4 =	36 ÷ 4 =	24 ÷ 4 =	40 ÷ 4 =
15 ÷ 3 =	27 ÷ 3 =	6 ÷ 3 =	3 ÷ 3 =	21 ÷ 3 =
14 ÷ 2 =	4 ÷ 2 =	20 ÷ 2 =	10 ÷ 2 =	8 ÷ 2 =
1 ÷ 1 =	4 ÷ 1 =	7 ÷ 1 =	2 ÷ 1 =	3 ÷ 1 =
4 ÷ 4 =	36 ÷ 4 =	12 ÷ 4 =	20 ÷ 4 =	8 ÷ 4 =
6 ÷ 3 =	15 ÷ 3 =	21 ÷ 3 =	27 ÷ 3 =	3 ÷ 3 =
4 ÷ 2 =	8 ÷ 2 =	16 ÷ 2 =	20 ÷ 2 =	18 ÷ 2 =
2 ÷ 1 =	5 ÷ 1 =	8 ÷ 1 =	1 ÷ 1 =	3 ÷ 1 =
24 ÷ 4 =	28 ÷ 4 =	20 ÷ 4 =	16 ÷ 4 =	32 ÷ 4 =
30 ÷ 3 =	27 ÷ 3 =	12 ÷ 3 =	18 ÷ 3 =	24 ÷ 3 =
36 ÷ 4 =	40 ÷ 4 =	32 ÷ 4 =	8 ÷ 4 =	28 ÷ 4 =
24 ÷ 3 =	15 ÷ 3 =	30 ÷ 3 =	21 ÷ 3 =	18 ÷ 3 =
3 ÷ 1 =	6 ÷ 1 =	9 ÷ 1 =	10 ÷ 1 =	4 ÷ 1 =
10 ÷ 2 =	12 ÷ 2 =	14 ÷ 2 =	0 ÷ 2 =	6 ÷ 2 =
12 ÷ 3 =	30 ÷ 3 =	18 ÷ 3 =	9 ÷ 3 =	24 ÷ 3 =
40 ÷ 4 =	0 ÷ 4 =	12 ÷ 4 =	24 ÷ 4 =	16 ÷ 4 =
0 ÷ 1 =	10 ÷ 1 =	7 ÷ 1 =	1 ÷ 1 =	5 ÷ 1 =
6 ÷ 2 =	16 ÷ 2 =	18 ÷ 2 =	20 ÷ 2 =	12 ÷ 2 =
15 ÷ 3 =	24 ÷ 3 =	30 ÷ 3 =	18 ÷ 3 =	0 ÷ 3 =
20 ÷ 4 =	24 ÷ 4 =	36 ÷ 4 =	40 ÷ 4 =	4 ÷ 4 =

Dividing 5

$5\overline{)5}$ $5\overline{)15}$ $5\overline{)30}$ $5\overline{)10}$ $5\overline{)35}$ $5\overline{)20}$ $5\overline{)40}$ $5\overline{)25}$ $5\overline{)50}$ $5\overline{)45}$

$5\overline{)25}$ $5\overline{)10}$ $5\overline{)30}$ $5\overline{)5}$ $5\overline{)35}$ $5\overline{)45}$ $5\overline{)15}$ $5\overline{)40}$ $5\overline{)20}$ $5\overline{)50}$

$5\overline{)10}$ $5\overline{)40}$ $5\overline{)5}$ $5\overline{)30}$ $5\overline{)15}$ $5\overline{)25}$ $5\overline{)45}$ $5\overline{)50}$ $5\overline{)35}$ $5\overline{)20}$

$5\overline{)20}$ $5\overline{)35}$ $5\overline{)25}$ $5\overline{)10}$ $5\overline{)5}$ $5\overline{)50}$ $5\overline{)40}$ $5\overline{)15}$ $5\overline{)45}$ $5\overline{)5}$

$5\overline{)30}$ $5\overline{)5}$ $5\overline{)10}$ $5\overline{)25}$ $5\overline{)45}$ $5\overline{)15}$ $5\overline{)50}$ $5\overline{)35}$ $5\overline{)20}$ $5\overline{)40}$

$5\overline{)40}$ $5\overline{)45}$ $5\overline{)50}$ $5\overline{)0}$ $5\overline{)30}$ $5\overline{)25}$ $5\overline{)15}$ $5\overline{)20}$ $5\overline{)5}$ $5\overline{)10}$

$5\overline{)45}$ $5\overline{)50}$ $5\overline{)25}$ $5\overline{)40}$ $5\overline{)20}$ $5\overline{)35}$ $5\overline{)10}$ $5\overline{)30}$ $5\overline{)15}$ $5\overline{)5}$

$5\overline{)50}$ $5\overline{)20}$ $5\overline{)40}$ $5\overline{)15}$ $5\overline{)45}$ $5\overline{)35}$ $5\overline{)5}$ $5\overline{)30}$ $5\overline{)10}$ $5\overline{)25}$

$5\overline{)0}$ $5\overline{)35}$ $5\overline{)50}$ $5\overline{)45}$ $5\overline{)15}$ $5\overline{)50}$ $5\overline{)20}$ $5\overline{)40}$ $5\overline{)25}$ $5\overline{)10}$

$5\overline{)15}$ $5\overline{)25}$ $5\overline{)35}$ $5\overline{)20}$ $5\overline{)10}$ $5\overline{)5}$ $5\overline{)25}$ $5\overline{)5}$ $5\overline{)30}$ $5\overline{)15}$

Dividing 5

5 ÷ 5 =	30 ÷ 5 =	35 ÷ 5 =	50 ÷ 5 =	40 ÷ 5 =
25 ÷ 5 =	30 ÷ 5 =	15 ÷ 5 =	15 ÷ 5 =	20 ÷ 5 =
10 ÷ 5 =	5 ÷ 5 =	5 ÷ 5 =	45 ÷ 5 =	35 ÷ 5 =
20 ÷ 5 =	25 ÷ 5 =	35 ÷ 5 =	40 ÷ 5 =	45 ÷ 5 =
30 ÷ 5 =	10 ÷ 5 =	45 ÷ 5 =	50 ÷ 5 =	20 ÷ 5 =
40 ÷ 5 =	50 ÷ 5 =	30 ÷ 5 =	15 ÷ 5 =	5 ÷ 5 =
45 ÷ 5 =	25 ÷ 5 =	20 ÷ 5 =	10 ÷ 5 =	15 ÷ 5 =
50 ÷ 5 =	40 ÷ 5 =	45 ÷ 5 =	5 ÷ 5 =	10 ÷ 5 =
0 ÷ 5 =	50 ÷ 5 =	15 ÷ 5 =	20 ÷ 5 =	25 ÷ 5 =
15 ÷ 5 =	35 ÷ 5 =	10 ÷ 5 =	25 ÷ 5 =	30 ÷ 5 =
25 ÷ 5 =	10 ÷ 5 =	5 ÷ 5 =	5 ÷ 5 =	45 ÷ 5 =
35 ÷ 5 =	5 ÷ 5 =	30 ÷ 5 =	40 ÷ 5 =	50 ÷ 5 =
20 ÷ 5 =	30 ÷ 5 =	35 ÷ 5 =	30 ÷ 5 =	20 ÷ 5 =
50 ÷ 5 =	10 ÷ 5 =	25 ÷ 5 =	20 ÷ 5 =	5 ÷ 5 =
45 ÷ 5 =	25 ÷ 5 =	15 ÷ 5 =	35 ÷ 5 =	40 ÷ 5 =
5 ÷ 5 =	0 ÷ 5 =	35 ÷ 5 =	30 ÷ 5 =	10 ÷ 5 =
35 ÷ 5 =	40 ÷ 5 =	50 ÷ 5 =	15 ÷ 5 =	5 ÷ 5 =
40 ÷ 5 =	15 ÷ 5 =	25 ÷ 5 =	50 ÷ 5 =	25 ÷ 5 =
10 ÷ 5 =	45 ÷ 5 =	45 ÷ 5 =	40 ÷ 5 =	10 ÷ 5 =
15 ÷ 5 =	20 ÷ 5 =	25 ÷ 5 =	20 ÷ 5 =	15 ÷ 5 =

Dividing 6

$6\overline{)6}$ $6\overline{)18}$ $6\overline{)12}$ $6\overline{)30}$ $6\overline{)54}$ $6\overline{)24}$ $6\overline{)60}$ $6\overline{)36}$ $6\overline{)48}$ $6\overline{)42}$

$6\overline{)18}$ $6\overline{)36}$ $6\overline{)48}$ $6\overline{)6}$ $6\overline{)30}$ $6\overline{)54}$ $6\overline{)42}$ $6\overline{)12}$ $6\overline{)60}$ $6\overline{)24}$

$6\overline{)0}$ $6\overline{)30}$ $6\overline{)12}$ $6\overline{)54}$ $6\overline{)6}$ $6\overline{)48}$ $6\overline{)24}$ $6\overline{)42}$ $6\overline{)18}$ $6\overline{)60}$

$6\overline{)30}$ $6\overline{)12}$ $6\overline{)54}$ $6\overline{)6}$ $6\overline{)48}$ $6\overline{)24}$ $6\overline{)60}$ $6\overline{)18}$ $6\overline{)36}$ $6\overline{)42}$

$6\overline{)24}$ $6\overline{)6}$ $6\overline{)42}$ $6\overline{)12}$ $6\overline{)60}$ $6\overline{)48}$ $6\overline{)18}$ $6\overline{)0}$ $6\overline{)30}$ $6\overline{)54}$

$6\overline{)60}$ $6\overline{)48}$ $6\overline{)54}$ $6\overline{)18}$ $6\overline{)36}$ $6\overline{)42}$ $6\overline{)30}$ $6\overline{)24}$ $6\overline{)12}$ $6\overline{)6}$

$6\overline{)48}$ $6\overline{)24}$ $6\overline{)6}$ $6\overline{)54}$ $6\overline{)12}$ $6\overline{)60}$ $6\overline{)36}$ $6\overline{)42}$ $6\overline{)18}$ $6\overline{)30}$

$6\overline{)36}$ $6\overline{)42}$ $6\overline{)18}$ $6\overline{)24}$ $6\overline{)60}$ $6\overline{)54}$ $6\overline{)48}$ $6\overline{)30}$ $6\overline{)6}$ $6\overline{)12}$

$6\overline{)12}$ $6\overline{)54}$ $6\overline{)30}$ $6\overline{)36}$ $6\overline{)42}$ $6\overline{)18}$ $6\overline{)6}$ $6\overline{)48}$ $6\overline{)24}$ $6\overline{)18}$

$6\overline{)42}$ $6\overline{)60}$ $6\overline{)24}$ $6\overline{)60}$ $6\overline{)18}$ $6\overline{)12}$ $6\overline{)54}$ $6\overline{)6}$ $6\overline{)42}$ $6\overline{)36}$

6s

Dividing 6

6 ÷ 6 =	12 ÷ 6 =	54 ÷ 6 =	60 ÷ 6 =	24 ÷ 6 =
18 ÷ 6 =	48 ÷ 6 =	30 ÷ 6 =	42 ÷ 6 =	60 ÷ 6 =
0 ÷ 6 =	12 ÷ 6 =	6 ÷ 6 =	24 ÷ 6 =	18 ÷ 6 =
30 ÷ 6 =	54 ÷ 6 =	48 ÷ 6 =	30 ÷ 6 =	36 ÷ 6 =
24 ÷ 6 =	42 ÷ 6 =	60 ÷ 6 =	18 ÷ 6 =	30 ÷ 6 =
60 ÷ 6 =	54 ÷ 6 =	36 ÷ 6 =	30 ÷ 6 =	12 ÷ 6 =
48 ÷ 6 =	6 ÷ 6 =	12 ÷ 6 =	36 ÷ 6 =	18 ÷ 6 =
36 ÷ 6 =	18 ÷ 6 =	60 ÷ 6 =	48 ÷ 6 =	6 ÷ 6 =
12 ÷ 6 =	30 ÷ 6 =	42 ÷ 6 =	6 ÷ 6 =	24 ÷ 6 =
42 ÷ 6 =	24 ÷ 6 =	18 ÷ 6 =	54 ÷ 6 =	42 ÷ 6 =
18 ÷ 6 =	60 ÷ 6 =	24 ÷ 6 =	36 ÷ 6 =	36 ÷ 6 =
36 ÷ 6 =	54 ÷ 6 =	36 ÷ 6 =	12 ÷ 6 =	18 ÷ 6 =
30 ÷ 6 =	24 ÷ 6 =	48 ÷ 6 =	42 ÷ 6 =	12 ÷ 6 =
12 ÷ 6 =	54 ÷ 6 =	24 ÷ 6 =	18 ÷ 6 =	30 ÷ 6 =
6 ÷ 6 =	18 ÷ 6 =	48 ÷ 6 =	0 ÷ 6 =	6 ÷ 6 =
48 ÷ 6 =	12 ÷ 6 =	42 ÷ 6 =	24 ÷ 6 =	54 ÷ 6 =
24 ÷ 6 =	6 ÷ 6 =	60 ÷ 6 =	42 ÷ 6 =	42 ÷ 6 =
42 ÷ 6 =	0 ÷ 6 =	54 ÷ 6 =	30 ÷ 6 =	60 ÷ 6 =
54 ÷ 6 =	30 ÷ 6 =	18 ÷ 6 =	48 ÷ 6 =	24 ÷ 6 =
60 ÷ 6 =	36 ÷ 6 =	12 ÷ 6 =	6 ÷ 6 =	42 ÷ 6 =

Dividing 7

$7\overline{)7}$ $7\overline{)14}$ $7\overline{)35}$ $7\overline{)21}$ $7\overline{)70}$ $7\overline{)63}$ $7\overline{)42}$ $7\overline{)49}$ $7\overline{)56}$ $7\overline{)28}$

$7\overline{)21}$ $7\overline{)42}$ $7\overline{)63}$ $7\overline{)7}$ $7\overline{)35}$ $7\overline{)70}$ $7\overline{)28}$ $7\overline{)49}$ $7\overline{)14}$ $7\overline{)56}$

$7\overline{)28}$ $7\overline{)49}$ $7\overline{)21}$ $7\overline{)63}$ $7\overline{)7}$ $7\overline{)35}$ $7\overline{)70}$ $7\overline{)14}$ $7\overline{)42}$ $7\overline{)0}$

$7\overline{)49}$ $7\overline{)21}$ $7\overline{)35}$ $7\overline{)7}$ $7\overline{)63}$ $7\overline{)42}$ $7\overline{)56}$ $7\overline{)70}$ $7\overline{)14}$ $7\overline{)28}$

$7\overline{)14}$ $7\overline{)35}$ $7\overline{)42}$ $7\overline{)49}$ $7\overline{)70}$ $7\overline{)7}$ $7\overline{)28}$ $7\overline{)56}$ $7\overline{)63}$ $7\overline{)21}$

$7\overline{)0}$ $7\overline{)42}$ $7\overline{)14}$ $7\overline{)70}$ $7\overline{)35}$ $7\overline{)7}$ $7\overline{)63}$ $7\overline{)21}$ $7\overline{)49}$ $7\overline{)28}$

$7\overline{)56}$ $7\overline{)28}$ $7\overline{)49}$ $7\overline{)42}$ $7\overline{)21}$ $7\overline{)70}$ $7\overline{)35}$ $7\overline{)14}$ $7\overline{)63}$ $7\overline{)7}$

$7\overline{)35}$ $7\overline{)7}$ $7\overline{)28}$ $7\overline{)14}$ $7\overline{)28}$ $7\overline{)49}$ $7\overline{)21}$ $7\overline{)7}$ $7\overline{)70}$ $7\overline{)14}$

$7\overline{)42}$ $7\overline{)49}$ $7\overline{)56}$ $7\overline{)28}$ $7\overline{)14}$ $7\overline{)21}$ $7\overline{)14}$ $7\overline{)35}$ $7\overline{)7}$ $7\overline{)35}$

$7\overline{)63}$ $7\overline{)56}$ $7\overline{)7}$ $7\overline{)35}$ $7\overline{)28}$ $7\overline{)14}$ $7\overline{)7}$ $7\overline{)42}$ $7\overline{)21}$ $7\overline{)70}$

Dividing 7

7 ÷ 7 =	35 ÷ 7 =	70 ÷ 7 =	42 ÷ 7 =	56 ÷ 7 =
21 ÷ 7 =	63 ÷ 7 =	35 ÷ 7 =	49 ÷ 7 =	28 ÷ 7 =
28 ÷ 7 =	21 ÷ 7 =	7 ÷ 7 =	70 ÷ 7 =	42 ÷ 7 =
49 ÷ 7 =	35 ÷ 7 =	63 ÷ 7 =	56 ÷ 7 =	14 ÷ 7 =
14 ÷ 7 =	42 ÷ 7 =	70 ÷ 7 =	28 ÷ 7 =	63 ÷ 7 =
0 ÷ 7 =	14 ÷ 7 =	35 ÷ 7 =	63 ÷ 7 =	49 ÷ 7 =
56 ÷ 7 =	49 ÷ 7 =	21 ÷ 7 =	35 ÷ 7 =	63 ÷ 7 =
35 ÷ 7 =	28 ÷ 7 =	28 ÷ 7 =	21 ÷ 7 =	70 ÷ 7 =
42 ÷ 7 =	56 ÷ 7 =	14 ÷ 7 =	14 ÷ 7 =	7 ÷ 7 =
63 ÷ 7 =	7 ÷ 7 =	28 ÷ 7 =	7 ÷ 7 =	21 ÷ 7 =
14 ÷ 7 =	21 ÷ 7 =	63 ÷ 7 =	49 ÷ 7 =	28 ÷ 7 =
42 ÷ 7 =	63 ÷ 7 =	70 ÷ 7 =	14 ÷ 7 =	56 ÷ 7 =
49 ÷ 7 =	7 ÷ 7 =	35 ÷ 7 =	70 ÷ 7 =	0 ÷ 7 =
21 ÷ 7 =	49 ÷ 7 =	42 ÷ 7 =	14 ÷ 7 =	28 ÷ 7 =
35 ÷ 7 =	70 ÷ 7 =	7 ÷ 7 =	56 ÷ 7 =	21 ÷ 7 =
42 ÷ 7 =	7 ÷ 7 =	21 ÷ 7 =	21 ÷ 7 =	28 ÷ 7 =
28 ÷ 7 =	42 ÷ 7 =	7 ÷ 7 =	14 ÷ 7 =	7 ÷ 7 =
7 ÷ 7 =	14 ÷ 7 =	49 ÷ 7 =	7 ÷ 7 =	14 ÷ 7 =
49 ÷ 7 =	28 ÷ 7 =	21 ÷ 7 =	35 ÷ 7 =	35 ÷ 7 =
56 ÷ 7 =	35 ÷ 7 =	14 ÷ 7 =	42 ÷ 7 =	70 ÷ 7 =

$5\overline{)10}$ $7\overline{)63}$ $6\overline{)36}$ $5\overline{)35}$ $7\overline{)49}$ $6\overline{)48}$ $5\overline{)40}$ $7\overline{)35}$ $6\overline{)42}$ $6\overline{)54}$

$6\overline{)18}$ $5\overline{)5}$ $7\overline{)56}$ $6\overline{)42}$ $5\overline{)30}$ $7\overline{)42}$ $6\overline{)42}$ $5\overline{)45}$ $6\overline{)6}$ $6\overline{)48}$

$7\overline{)70}$ $6\overline{)12}$ $5\overline{)10}$ $7\overline{)49}$ $6\overline{)48}$ $5\overline{)25}$ $7\overline{)21}$ $6\overline{)36}$ $5\overline{)50}$ $7\overline{)49}$

$4\overline{)40}$ $7\overline{)7}$ $6\overline{)6}$ $5\overline{)15}$ $7\overline{)42}$ $6\overline{)54}$ $5\overline{)20}$ $7\overline{)70}$ $6\overline{)54}$ $5\overline{)5}$

$6\overline{)60}$ $5\overline{)45}$ $7\overline{)14}$ $6\overline{)60}$ $5\overline{)20}$ $7\overline{)35}$ $6\overline{)60}$ $5\overline{)15}$ $7\overline{)63}$ $6\overline{)18}$

$7\overline{)14}$ $6\overline{)6}$ $5\overline{)40}$ $7\overline{)21}$ $6\overline{)48}$ $5\overline{)25}$ $7\overline{)28}$ $6\overline{)18}$ $5\overline{)10}$ $7\overline{)56}$

$5\overline{)50}$ $7\overline{)7}$ $6\overline{)12}$ $5\overline{)35}$ $4\overline{)28}$ $6\overline{)42}$ $6\overline{)30}$ $7\overline{)21}$ $6\overline{)12}$ $5\overline{)5}$

$6\overline{)48}$ $5\overline{)5}$ $7\overline{)70}$ $6\overline{)18}$ $5\overline{)30}$ $7\overline{)35}$ $6\overline{)36}$ $5\overline{)35}$ $7\overline{)14}$ $6\overline{)18}$

$7\overline{)21}$ $6\overline{)42}$ $5\overline{)10}$ $7\overline{)63}$ $6\overline{)24}$ $5\overline{)30}$ $7\overline{)42}$ $6\overline{)30}$ $5\overline{)40}$ $7\overline{)7}$

$5\overline{)35}$ $7\overline{)28}$ $6\overline{)36}$ $5\overline{)15}$ $7\overline{)56}$ $6\overline{)30}$ $5\overline{)25}$ $7\overline{)49}$ $6\overline{)24}$ $5\overline{)45}$

Section Diagnostic Test 5 – 7

20 ÷ 5 =	25 ÷ 5 =	35 ÷ 5 =	45 ÷ 5 =	40 ÷ 5 =
42 ÷ 6 =	54 ÷ 6 =	48 ÷ 6 =	30 ÷ 6 =	36 ÷ 6 =
7 ÷ 7 =	21 ÷ 7 =	14 ÷ 7 =	28 ÷ 7 =	70 ÷ 7 =
15 ÷ 5 =	5 ÷ 5 =	50 ÷ 5 =	10 ÷ 5 =	30 ÷ 5 =
6 ÷ 6 =	60 ÷ 6 =	24 ÷ 6 =	18 ÷ 6 =	12 ÷ 6 =
35 ÷ 7 =	49 ÷ 7 =	63 ÷ 7 =	56 ÷ 7 =	42 ÷ 7 =
10 ÷ 5 =	20 ÷ 5 =	30 ÷ 5 =	40 ÷ 5 =	50 ÷ 5 =
6 ÷ 6 =	18 ÷ 6 =	54 ÷ 6 =	30 ÷ 6 =	42 ÷ 6 =
63 ÷ 7 =	49 ÷ 7 =	35 ÷ 7 =	21 ÷ 7 =	7 ÷ 7 =
15 ÷ 5 =	25 ÷ 5 =	35 ÷ 5 =	45 ÷ 5 =	5 ÷ 5 =
12 ÷ 6 =	60 ÷ 6 =	36 ÷ 6 =	24 ÷ 6 =	48 ÷ 6 =
56 ÷ 7 =	42 ÷ 7 =	28 ÷ 7 =	14 ÷ 7 =	70 ÷ 7 =
45 ÷ 5 =	35 ÷ 5 =	40 ÷ 5 =	30 ÷ 5 =	25 ÷ 5 =
36 ÷ 6 =	12 ÷ 6 =	54 ÷ 6 =	60 ÷ 6 =	42 ÷ 6 =
49 ÷ 7 =	21 ÷ 7 =	42 ÷ 7 =	35 ÷ 7 =	70 ÷ 7 =
20 ÷ 5 =	50 ÷ 5 =	10 ÷ 5 =	5 ÷ 5 =	15 ÷ 5 =
18 ÷ 6 =	48 ÷ 6 =	30 ÷ 6 =	24 ÷ 6 =	6 ÷ 6 =
63 ÷ 7 =	28 ÷ 7 =	56 ÷ 7 =	7 ÷ 7 =	14 ÷ 7 =
60 ÷ 6 =	48 ÷ 6 =	54 ÷ 6 =	42 ÷ 6 =	24 ÷ 6 =
42 ÷ 7 =	49 ÷ 7 =	24 ÷ 7 =	56 ÷ 7 =	63 ÷ 7 =

Dividing 8

$8\overline{)8}$ $8\overline{)24}$ $8\overline{)16}$ $8\overline{)32}$ $8\overline{)48}$ $8\overline{)72}$ $8\overline{)80}$ $8\overline{)56}$ $8\overline{)64}$ $8\overline{)40}$

$8\overline{)24}$ $8\overline{)56}$ $8\overline{)40}$ $8\overline{)16}$ $8\overline{)72}$ $8\overline{)48}$ $8\overline{)8}$ $8\overline{)64}$ $8\overline{)32}$ $8\overline{)80}$

$8\overline{)32}$ $8\overline{)64}$ $8\overline{)8}$ $8\overline{)40}$ $8\overline{)48}$ $8\overline{)72}$ $8\overline{)16}$ $8\overline{)56}$ $8\overline{)80}$ $8\overline{)24}$

$8\overline{)40}$ $8\overline{)80}$ $8\overline{)64}$ $8\overline{)32}$ $8\overline{)8}$ $8\overline{)24}$ $8\overline{)48}$ $8\overline{)16}$ $8\overline{)56}$ $8\overline{)72}$

$8\overline{)16}$ $8\overline{)24}$ $8\overline{)80}$ $8\overline{)72}$ $8\overline{)32}$ $8\overline{)64}$ $8\overline{)56}$ $8\overline{)8}$ $8\overline{)40}$ $8\overline{)48}$

$8\overline{)80}$ $8\overline{)32}$ $8\overline{)8}$ $8\overline{)64}$ $8\overline{)72}$ $8\overline{)48}$ $8\overline{)40}$ $8\overline{)16}$ $8\overline{)24}$ $8\overline{)56}$

$8\overline{)64}$ $8\overline{)40}$ $8\overline{)24}$ $8\overline{)48}$ $8\overline{)40}$ $8\overline{)72}$ $8\overline{)8}$ $8\overline{)80}$ $8\overline{)16}$ $8\overline{)8}$

$8\overline{)8}$ $8\overline{)72}$ $8\overline{)32}$ $8\overline{)8}$ $8\overline{)56}$ $8\overline{)16}$ $8\overline{)24}$ $8\overline{)72}$ $8\overline{)24}$ $8\overline{)16}$

$8\overline{)72}$ $8\overline{)48}$ $8\overline{)24}$ $8\overline{)80}$ $8\overline{)16}$ $8\overline{)32}$ $8\overline{)24}$ $8\overline{)40}$ $8\overline{)8}$ $8\overline{)32}$

$8\overline{)48}$ $8\overline{)8}$ $8\overline{)56}$ $8\overline{)24}$ $8\overline{)64}$ $8\overline{)80}$ $8\overline{)32}$ $8\overline{)24}$ $8\overline{)56}$ $8\overline{)72}$

8s

Dividing 8

8 ÷ 8 =	16 ÷ 8 =	48 ÷ 8 =	80 ÷ 8 =	64 ÷ 8 =
24 ÷ 8 =	40 ÷ 8 =	72 ÷ 8 =	8 ÷ 8 =	32 ÷ 8 =
32 ÷ 8 =	8 ÷ 8 =	48 ÷ 8 =	16 ÷ 8 =	80 ÷ 8 =
40 ÷ 8 =	64 ÷ 8 =	8 ÷ 8 =	0 ÷ 8 =	56 ÷ 8 =
16 ÷ 8 =	80 ÷ 8 =	32 ÷ 8 =	56 ÷ 8 =	40 ÷ 8 =
80 ÷ 8 =	8 ÷ 8 =	72 ÷ 8 =	40 ÷ 8 =	24 ÷ 8 =
64 ÷ 8 =	24 ÷ 8 =	40 ÷ 8 =	8 ÷ 8 =	16 ÷ 8 =
8 ÷ 8 =	32 ÷ 8 =	56 ÷ 8 =	24 ÷ 8 =	24 ÷ 8 =
72 ÷ 8 =	24 ÷ 8 =	16 ÷ 8 =	0 ÷ 8 =	8 ÷ 8 =
48 ÷ 8 =	56 ÷ 8 =	64 ÷ 8 =	32 ÷ 8 =	56 ÷ 8 =
56 ÷ 8 =	72 ÷ 8 =	24 ÷ 8 =	72 ÷ 8 =	80 ÷ 8 =
24 ÷ 8 =	32 ÷ 8 =	72 ÷ 8 =	56 ÷ 8 =	40 ÷ 8 =
64 ÷ 8 =	16 ÷ 8 =	48 ÷ 8 =	64 ÷ 8 =	80 ÷ 8 =
56 ÷ 8 =	40 ÷ 8 =	72 ÷ 8 =	56 ÷ 8 =	24 ÷ 8 =
80 ÷ 8 =	32 ÷ 8 =	24 ÷ 8 =	16 ÷ 8 =	72 ÷ 8 =
24 ÷ 8 =	72 ÷ 8 =	64 ÷ 8 =	8 ÷ 8 =	48 ÷ 8 =
32 ÷ 8 =	64 ÷ 8 =	48 ÷ 8 =	16 ÷ 8 =	56 ÷ 8 =
40 ÷ 8 =	48 ÷ 8 =	72 ÷ 8 =	80 ÷ 8 =	8 ÷ 8 =
72 ÷ 8 =	8 ÷ 8 =	16 ÷ 8 =	72 ÷ 8 =	16 ÷ 8 =
48 ÷ 8 =	80 ÷ 8 =	32 ÷ 8 =	40 ÷ 8 =	32 ÷ 8 =

Dividing 9

$9\overline{)9}$ $9\overline{)18}$ $9\overline{)36}$ $9\overline{)90}$ $9\overline{)27}$ $9\overline{)45}$ $9\overline{)81}$ $9\overline{)63}$ $9\overline{)72}$ $9\overline{)54}$

$9\overline{)18}$ $9\overline{)90}$ $9\overline{)63}$ $9\overline{)27}$ $9\overline{)9}$ $9\overline{)54}$ $9\overline{)72}$ $9\overline{)36}$ $9\overline{)81}$ $9\overline{)45}$

$9\overline{)45}$ $9\overline{)81}$ $9\overline{)36}$ $9\overline{)72}$ $9\overline{)54}$ $9\overline{)9}$ $9\overline{)27}$ $9\overline{)63}$ $9\overline{)90}$ $9\overline{)18}$

$9\overline{)27}$ $9\overline{)54}$ $9\overline{)81}$ $9\overline{)36}$ $9\overline{)90}$ $9\overline{)72}$ $9\overline{)9}$ $9\overline{)54}$ $9\overline{)18}$ $9\overline{)63}$

$9\overline{)63}$ $9\overline{)18}$ $9\overline{)54}$ $9\overline{)9}$ $9\overline{)72}$ $9\overline{)90}$ $9\overline{)36}$ $9\overline{)81}$ $9\overline{)45}$ $9\overline{)27}$

$9\overline{)36}$ $9\overline{)54}$ $9\overline{)18}$ $9\overline{)72}$ $9\overline{)81}$ $9\overline{)63}$ $9\overline{)90}$ $9\overline{)27}$ $9\overline{)9}$ $9\overline{)45}$

$9\overline{)45}$ $9\overline{)9}$ $9\overline{)27}$ $9\overline{)90}$ $9\overline{)63}$ $9\overline{)81}$ $9\overline{)72}$ $9\overline{)18}$ $9\overline{)54}$ $9\overline{)36}$

$9\overline{)90}$ $9\overline{)27}$ $9\overline{)9}$ $9\overline{)81}$ $9\overline{)45}$ $9\overline{)18}$ $9\overline{)36}$ $9\overline{)90}$ $9\overline{)27}$ $9\overline{)72}$

$9\overline{)54}$ $9\overline{)36}$ $9\overline{)45}$ $9\overline{)18}$ $9\overline{)36}$ $9\overline{)27}$ $9\overline{)54}$ $9\overline{)45}$ $9\overline{)18}$ $9\overline{)9}$

$9\overline{)72}$ $9\overline{)63}$ $9\overline{)90}$ $9\overline{)54}$ $9\overline{)18}$ $9\overline{)36}$ $9\overline{)45}$ $9\overline{)72}$ $9\overline{)9}$ $9\overline{)0}$

Dividing 9

9 ÷ 9 =	36 ÷ 9 =	27 ÷ 9 =	81 ÷ 9 =	72 ÷ 9 =
18 ÷ 9 =	63 ÷ 9 =	9 ÷ 9 =	72 ÷ 9 =	81 ÷ 9 =
45 ÷ 9 =	81 ÷ 9 =	54 ÷ 9 =	27 ÷ 9 =	90 ÷ 9 =
27 ÷ 9 =	36 ÷ 9 =	90 ÷ 9 =	9 ÷ 9 =	18 ÷ 9 =
63 ÷ 9 =	18 ÷ 9 =	72 ÷ 9 =	36 ÷ 9 =	45 ÷ 9 =
36 ÷ 9 =	27 ÷ 9 =	81 ÷ 9 =	90 ÷ 9 =	9 ÷ 9 =
45 ÷ 9 =	9 ÷ 9 =	63 ÷ 9 =	72 ÷ 9 =	54 ÷ 9 =
90 ÷ 9 =	45 ÷ 9 =	0 ÷ 9 =	18 ÷ 9 =	27 ÷ 9 =
72 ÷ 9 =	54 ÷ 9 =	18 ÷ 9 =	45 ÷ 9 =	72 ÷ 9 =
18 ÷ 9 =	18 ÷ 9 =	45 ÷ 9 =	63 ÷ 9 =	0 ÷ 9 =
90 ÷ 9 =	45 ÷ 9 =	54 ÷ 9 =	36 ÷ 9 =	9 ÷ 9 =
81 ÷ 9 =	90 ÷ 9 =	9 ÷ 9 =	63 ÷ 9 =	72 ÷ 9 =
45 ÷ 9 =	72 ÷ 9 =	72 ÷ 9 =	54 ÷ 9 =	36 ÷ 9 =
18 ÷ 9 =	9 ÷ 9 =	90 ÷ 9 =	81 ÷ 9 =	45 ÷ 9 =
54 ÷ 9 =	36 ÷ 9 =	63 ÷ 9 =	27 ÷ 9 =	27 ÷ 9 =
9 ÷ 9 =	72 ÷ 9 =	81 ÷ 9 =	18 ÷ 9 =	63 ÷ 9 =
27 ÷ 9 =	0 ÷ 9 =	18 ÷ 9 =	90 ÷ 9 =	18 ÷ 9 =
36 ÷ 9 =	90 ÷ 9 =	27 ÷ 9 =	45 ÷ 9 =	45 ÷ 9 =
63 ÷ 9 =	54 ÷ 9 =	36 ÷ 9 =	9 ÷ 9 =	54 ÷ 9 =
54 ÷ 9 =	90 ÷ 9 =	36 ÷ 9 =	54 ÷ 9 =	18 ÷ 9 =

Dividing 10

10)100 10)80 10)30 10)90 10)40 10)10 10)60 10)20 10)50 10)70

10)60 10)10 10)90 10)50 10)80 10)30 10)70 10)100 10)40 10)20

10)70 10)50 10)20 10)60 10)10 10)40 10)90 10)30 10)80 10)100

10)60 10)20 10)100 10)70 10)30 10)80 10)50 10)90 10)10 10)40

10)50 10)80 10)30 10)20 10)70 10)90 10)60 10)40 10)100 10)10

10)90 10)60 10)20 10)40 10)90 10)100 10)80 10)10 10)50 10)30

10)10 10)100 10)40 10)90 10)60 10)70 10)20 10)30 10)80 10)50

10)30 10)50 10)10 10)80 10)100 10)20 10)90 10)40 10)60 10)80

10)80 10)40 10)30 10)20 10)10 10)50 10)100 10)60 10)90 10)70

10)70 10)90 10)80 10)100 10)50 10)60 10)10 10)20 10)30 10)40

Dividing 10

10 ÷ 10 =	20 ÷ 10 =	70 ÷ 10 =	80 ÷ 10 =	90 ÷ 10 =
30 ÷ 10 =	100 ÷ 10 =	60 ÷ 10 =	10 ÷ 10 =	50 ÷ 10 =
40 ÷ 10 =	30 ÷ 10 =	100 ÷ 10 =	50 ÷ 10 =	60 ÷ 10 =
80 ÷ 10 =	90 ÷ 10 =	40 ÷ 10 =	20 ÷ 10 =	70 ÷ 10 =
90 ÷ 10 =	40 ÷ 10 =	10 ÷ 10 =	80 ÷ 10 =	20 ÷ 10 =
100 ÷ 10 =	10 ÷ 10 =	30 ÷ 10 =	60 ÷ 10 =	40 ÷ 10 =
70 ÷ 10 =	30 ÷ 10 =	50 ÷ 10 =	100 ÷ 10 =	90 ÷ 10 =
20 ÷ 10 =	40 ÷ 10 =	90 ÷ 10 =	50 ÷ 10 =	80 ÷ 10 =
50 ÷ 10 =	60 ÷ 10 =	70 ÷ 10 =	40 ÷ 10 =	20 ÷ 10 =
60 ÷ 10 =	20 ÷ 10 =	40 ÷ 10 =	90 ÷ 10 =	100 ÷ 10 =
10 ÷ 10 =	50 ÷ 10 =	100 ÷ 10 =	30 ÷ 10 =	40 ÷ 10 =
100 ÷ 10 =	20 ÷ 10 =	40 ÷ 10 =	90 ÷ 10 =	80 ÷ 10 =
90 ÷ 10 =	80 ÷ 10 =	70 ÷ 10 =	20 ÷ 10 =	10 ÷ 10 =
20 ÷ 10 =	10 ÷ 10 =	60 ÷ 10 =	100 ÷ 10 =	30 ÷ 10 =
80 ÷ 10 =	100 ÷ 10 =	50 ÷ 10 =	30 ÷ 10 =	70 ÷ 10 =
60 ÷ 10 =	50 ÷ 10 =	90 ÷ 10 =	20 ÷ 10 =	90 ÷ 10 =
50 ÷ 10 =	80 ÷ 10 =	10 ÷ 10 =	40 ÷ 10 =	60 ÷ 10 =
90 ÷ 10 =	60 ÷ 10 =	30 ÷ 10 =	10 ÷ 10 =	100 ÷ 10 =
70 ÷ 10 =	90 ÷ 10 =	80 ÷ 10 =	30 ÷ 10 =	10 ÷ 10 =
60 ÷ 10 =	30 ÷ 10 =	70 ÷ 10 =	80 ÷ 10 =	50 ÷ 10 =

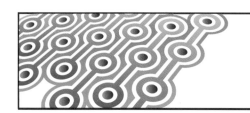

Review Sheet 8 – 10

8 ÷ 8 =	20 ÷ 10 =	72 ÷ 9 =	72 ÷ 8 =	18 ÷ 9 =
81 ÷ 9 =	80 ÷ 8 =	10 ÷ 10 =	36 ÷ 9 =	100 ÷ 10 =
100 ÷ 10 =	27 ÷ 9 =	8 ÷ 8 =	30 ÷ 10 =	72 ÷ 8 =
24 ÷ 8 =	48 ÷ 8 =	63 ÷ 9 =	64 ÷ 8 =	63 ÷ 9 =
40 ÷ 10 =	27 ÷ 9 =	32 ÷ 8 =	45 ÷ 9 =	30 ÷ 10 =
45 ÷ 9 =	16 ÷ 8 =	54 ÷ 9 =	60 ÷ 10 =	56 ÷ 8 =
24 ÷ 8 =	36 ÷ 9 =	60 ÷ 10 =	40 ÷ 8 =	63 ÷ 9 =
40 ÷ 10 =	32 ÷ 8 =	27 ÷ 9 =	36 ÷ 9 =	60 ÷ 10 =
48 ÷ 8 =	64 ÷ 8 =	30 ÷ 10 =	56 ÷ 8 =	48 ÷ 8 =
9 ÷ 9 =	27 ÷ 9 =	90 ÷ 9 =	9 ÷ 9 =	18 ÷ 9 =
90 ÷ 10 =	18 ÷ 9 =	40 ÷ 8 =	72 ÷ 8 =	48 ÷ 8 =
54 ÷ 9 =	40 ÷ 8 =	72 ÷ 9 =	50 ÷ 10 =	70 ÷ 10 =
64 ÷ 8 =	10 ÷ 10 =	32 ÷ 8 =	36 ÷ 9 =	72 ÷ 9 =
100 ÷ 10 =	48 ÷ 8 =	80 ÷ 10 =	70 ÷ 10 =	54 ÷ 9 =
32 ÷ 8 =	45 ÷ 9 =	24 ÷ 8 =	81 ÷ 9 =	16 ÷ 8 =
27 ÷ 9 =	56 ÷ 8 =	72 ÷ 9 =	64 ÷ 8 =	100 ÷ 10 =
64 ÷ 8 =	63 ÷ 9 =	48 ÷ 8 =	90 ÷ 10 =	54 ÷ 9 =
40 ÷ 10 =	72 ÷ 8 =	80 ÷ 10 =	8 ÷ 8 =	50 ÷ 10 =
63 ÷ 9 =	80 ÷ 10 =	40 ÷ 10 =	81 ÷ 9 =	80 ÷ 8 =
80 ÷ 8 =	63 ÷ 9 =	90 ÷ 9 =	10 ÷ 10 =	72 ÷ 9 =

Section Diagnostic Test 8 – 10

9)9 8)64 10)100 8)48 9)90 10)20 8)56 9)36 10)100 8)48

9)81 8)48 10)20 8)64 9)81 10)40 8)8 9)72 10)70 8)24

9)63 8)32 10)50 8)56 9)72 10)60 8)80 9)81 10)50 8)56

9)45 8)16 10)60 8)40 9)63 10)80 8)24 9)63 10)20 8)32

9)27 8)80 10)80 8)64 9)81 10)100 8)40 9)54 10)40 8)64

9)18 8)8 10)90 8)72 9)45 10)10 8)48 9)63 10)80 8)72

9)36 8)24 10)70 8)24 9)36 10)30 8)72 9)36 10)60 8)56

9)54 8)40 10)40 8)32 9)27 10)50 8)64 9)54 10)30 8)72

9)72 8)56 10)30 8)48 9)18 10)70 8)32 9)81 10)90 8)40

9)90 8)72 10)10 8)32 9)9 10)90 8)16 9)45 10)70 8)64

14 ÷ 7 =	6 ÷ 1 =	24 ÷ 4 =	15 ÷ 3 =	30 ÷ 3 =
54 ÷ 9 =	63 ÷ 9 =	36 ÷ 6 =	14 ÷ 2 =	4 ÷ 1 =
42 ÷ 7 =	15 ÷ 5 =	45 ÷ 5 =	27 ÷ 9 =	20 ÷ 4 =
64 ÷ 8 =	24 ÷ 4 =	12 ÷ 4 =	35 ÷ 7 =	18 ÷ 2 =
42 ÷ 7 =	90 ÷ 9 =	60 ÷ 6 =	48 ÷ 6 =	42 ÷ 6 =
81 ÷ 9 =	48 ÷ 8 =	21 ÷ 3 =	20 ÷ 2 =	32 ÷ 4 =
32 ÷ 8 =	5 ÷ 1 =	45 ÷ 5 =	24 ÷ 3 =	4 ÷ 2 =
8 ÷ 2 =	16 ÷ 4 =	56 ÷ 7 =	48 ÷ 8 =	100 ÷ 10 =
63 ÷ 7 =	12 ÷ 6 =	27 ÷ 3 =	40 ÷ 4 =	8 ÷ 4 =
4 ÷ 2 =	54 ÷ 9 =	81 ÷ 9 =	70 ÷ 10 =	72 ÷ 9 =
3 ÷ 3 =	16 ÷ 2 =	48 ÷ 8 =	16 ÷ 2 =	25 ÷ 5 =
36 ÷ 4 =	12 ÷ 3 =	21 ÷ 7 =	4 ÷ 4 =	18 ÷ 2 =
12 ÷ 2 =	72 ÷ 8 =	54 ÷ 9 =	28 ÷ 7 =	16 ÷ 4 =
18 ÷ 6 =	56 ÷ 8 =	80 ÷ 8 =	18 ÷ 3 =	40 ÷ 5 =
49 ÷ 7 =	10 ÷ 2 =	28 ÷ 4 =	36 ÷ 9 =	72 ÷ 9 =
35 ÷ 5 =	63 ÷ 7 =	40 ÷ 8 =	42 ÷ 6 =	9 ÷ 3 =
30 ÷ 5 =	72 ÷ 9 =	6 ÷ 2 =	49 ÷ 7 =	36 ÷ 4 =
54 ÷ 9 =	50 ÷ 5 =	72 ÷ 9 =	12 ÷ 6 =	1 ÷ 1 =
9 ÷ 3 =	24 ÷ 8 =	24 ÷ 6 =	10 ÷ 5 =	63 ÷ 7 =
42 ÷ 7 =	30 ÷ 6 =	6 ÷ 3 =	70 ÷ 7 =	16 ÷ 8 =

Final Assessment Test 1 – 10

$9\overline{)18}$ $8\overline{)16}$ $7\overline{)7}$ $6\overline{)60}$ $5\overline{)20}$ $4\overline{)16}$ $5\overline{)5}$ $6\overline{)18}$ $7\overline{)28}$ $10\overline{)100}$

$3\overline{)12}$ $9\overline{)54}$ $8\overline{)48}$ $7\overline{)35}$ $6\overline{)12}$ $5\overline{)30}$ $4\overline{)28}$ $5\overline{)10}$ $6\overline{)36}$ $7\overline{)21}$

$4\overline{)8}$ $3\overline{)3}$ $9\overline{)72}$ $8\overline{)72}$ $7\overline{)14}$ $6\overline{)54}$ $5\overline{)45}$ $4\overline{)20}$ $5\overline{)50}$ $6\overline{)48}$

$2\overline{)6}$ $4\overline{)12}$ $3\overline{)9}$ $9\overline{)27}$ $8\overline{)64}$ $7\overline{)63}$ $6\overline{)30}$ $5\overline{)15}$ $4\overline{)36}$ $5\overline{)40}$

$1\overline{)3}$ $2\overline{)4}$ $4\overline{)40}$ $3\overline{)30}$ $9\overline{)81}$ $8\overline{)32}$ $7\overline{)70}$ $6\overline{)48}$ $5\overline{)35}$ $4\overline{)24}$

$3\overline{)27}$ $1\overline{)4}$ $2\overline{)2}$ $4\overline{)32}$ $3\overline{)15}$ $9\overline{)36}$ $8\overline{)0}$ $7\overline{)42}$ $6\overline{)24}$ $5\overline{)10}$

$8\overline{)8}$ $3\overline{)18}$ $1\overline{)7}$ $2\overline{)8}$ $4\overline{)12}$ $3\overline{)27}$ $9\overline{)18}$ $8\overline{)40}$ $7\overline{)56}$ $6\overline{)42}$

$2\overline{)18}$ $8\overline{)56}$ $3\overline{)30}$ $1\overline{)5}$ $2\overline{)10}$ $4\overline{)36}$ $3\overline{)18}$ $9\overline{)9}$ $8\overline{)80}$ $7\overline{)49}$

$10\overline{)60}$ $2\overline{)20}$ $8\overline{)72}$ $3\overline{)21}$ $1\overline{)6}$ $2\overline{)16}$ $4\overline{)24}$ $3\overline{)21}$ $9\overline{)18}$ $8\overline{)24}$

$10\overline{)80}$ $10\overline{)30}$ $2\overline{)0}$ $8\overline{)48}$ $3\overline{)0}$ $1\overline{)8}$ $2\overline{)12}$ $4\overline{)28}$ $3\overline{)9}$ $9\overline{)90}$

ANSWERS

The **Beginning Assessment Test** has facts arranged diagonally. This diagonal arrangement quickly identifies facts that are firm and facts that need attention.

Begin **Practice Sheets** at the level where several errors occur in fact diagonals. That may be with 2s for younger learners and 4s with older learners.

Beginning Assessment Test, page 7.

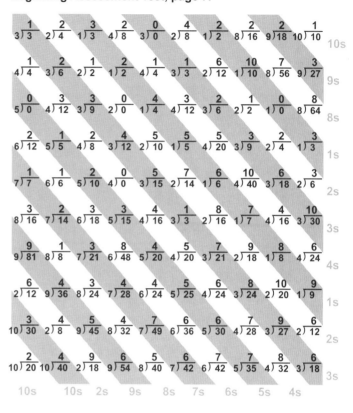

Practice Sheet 1s & 2s, page 8

1	1	2	2	3	3	4	4	5	5
6	6	7	7	8	0	9	9	10	10
5	5	4	4	3	3	2	1	2	7
9	1	6	8	6	7	10	10	9	0
10	8	3	4	5	5	0	3	7	7
2	6	0	1	10	5	4	2	6	8
10	7	9	8	8	9	7	10	6	10
10	1	8	2	3	3	2	2	1	0
2	0	3	9	4	8	5	7	6	6
7	5	8	4	9	3	10	2	0	1

Page 9

1	2	3	4	5
5	4	3	2	2
6	7	8	9	10
9	6	6	10	9
10	3	5	9	7
2	0	10	4	6
7	9	8	7	6
10	8	3	2	1
2	3	4	2	6
7	8	9	10	0
5	2	3	4	5
6	7	0	9	10
5	4	6	2	1
8	8	7	5	0
8	4	5	3	7
6	1	3	2	8
7	8	9	10	5
3	2	5	1	0
0	9	8	7	6
1	6	0	5	7

Practice Sheet 3s, page 10

1	2	0	3	4	6	5	8	9	7
6	8	2	9	1	4	5	3	7	10
5	1	8	10	4	2	6	9	7	3
2	5	4	7	1	10	6	8	3	0
9	5	1	4	7	3	8	10	2	9
3	4	7	1	8	10	2	3	9	5
6	7	2	10	5	8	1	9	4	3
2	10	5	7	1	6	8	3	9	4
4	8	3	6	10	7	1	9	5	2
0	2	5	9	4	8	10	1	3	6

Page 11

1	0	4	5	9
6	2	1	6	7
5	8	4	8	3
2	4	1	5	2
9	1	7	2	7
3	7	8	6	9
6	2	5	1	4
2	5	1	8	5
4	3	10	1	3
0	5	4	10	9
2	9	8	8	6
8	6	7	3	2
1	7	6	9	4
5	10	8	8	3
4	1	10	10	5
5	4	3	3	1
7	7	0	9	0
10	10	2	3	3
8	9	4	9	10
2	3	6	1	8

Practice Sheet 4s, page 12

1	3	6	5	10	7	4	8	9	2
4	1	8	3	6	10	2	9	7	5
7	4	2	8	5	1	9	6	10	3
3	1	2	7	4	8	5	0	3	9
8	9	10	9	8	10	9	8	10	8
3	8	4	1	6	2	4	10	9	5
2	7	5	9	1	3	10	4	6	0
10	1	9	2	8	4	6	5	4	7
9	0	3	10	4	5	3	6	2	1
5	3	10	9	1	6	4	1	8	4

Page 13

1	6	10	4	9
4	8	6	2	7
7	2	5	9	10
3	2	4	5	3
8	10	8	9	10
3	4	6	4	9
2	5	1	10	6
10	9	8	6	4
9	3	4	3	2
5	10	1	4	8
3	5	7	8	2
1	3	10	9	5
4	8	1	6	3
1	7	8	0	9
9	9	1	8	8
7	1	2	1	5
1	9	4	5	0
0	2	5	6	7
3	10	6	2	1
8	6	0	1	4

The *Section Diagnostic Tests* are specially arranged, too. The arrangement helps to identify if there are still problems and which facts those problems are.

Review Sheet 1–4, page 14

10	4	6	3	10	4	7	9	1	9
8	0	3	8	6	5	4	8	2	8
3	8	10	6	4	8	2	7	3	0
6	10	4	1	3	7	10	6	9	6
2	10	2	4	1	3	8	10	4	5
8	10	9	9	8	1	3	0	7	5
7	9	1	10	7	6	5	2	8	6
9	7	8	7	5	9	0	3	6	7
6	8	5	5	2	10	6	4	8	8
5	10	7	2	6	1	9	5	10	9

Practice Sheet 5s, page 16

1	3	6	2	7	4	8	5	10	9
5	2	6	1	7	9	3	8	4	10
2	8	1	6	3	5	9	10	7	4
4	7	5	2	1	10	8	3	9	1
6	1	2	5	9	3	10	7	4	8
8	9	10	0	6	5	3	4	1	2
9	10	5	8	4	7	2	6	3	1
10	4	8	3	9	7	1	6	2	5
0	7	10	9	3	10	4	8	5	2
3	5	7	4	2	1	5	1	6	3

Page 17

1	6	7	10	8
5	6	3	3	4
2	1	1	9	7
4	5	7	8	9
6	2	9	10	4
8	10	6	3	1
9	5	4	2	3
10	8	9	1	2
0	10	3	4	5
3	7	2	5	6
5	2	1	1	9
7	1	6	8	10
4	6	7	6	4
10	2	5	4	1
9	5	3	7	8
1	0	7	6	2
7	8	10	3	1
8	3	5	10	5
2	9	9	8	2
3	4	5	4	3

Section Diagnostic Test 1–4, page 15

$20 \div 4 = 5$	$24 \div 4 = 6$	$36 \div 4 = 9$	$24 \div 4 = 6$	$40 \div 4 = 10$	4s
$15 \div 3 = 5$	$27 \div 3 = 9$	$6 \div 3 = 2$	$3 \div 3 = 1$	$21 \div 3 = 7$	3s
$14 \div 2 = 7$	$4 \div 2 = 2$	$20 \div 2 = 10$	$10 \div 2 = 5$	$8 \div 2 = 4$	2s
$1 \div 1 = 1$	$4 \div 1 = 4$	$7 \div 1 = 7$	$2 \div 1 = 2$	$3 \div 1 = 3$	1s
$4 \div 4 = 1$	$36 \div 4 = 9$	$12 \div 4 = 3$	$20 \div 4 = 5$	$8 \div 4 = 2$	4s
$6 \div 3 = 2$	$15 \div 3 = 5$	$21 \div 3 = 7$	$27 \div 3 = 9$	$3 \div 3 = 1$	3s
$4 \div 2 = 2$	$8 \div 2 = 4$	$16 \div 2 = 8$	$20 \div 2 = 10$	$18 \div 2 = 9$	2s
$2 \div 1 = 2$	$5 \div 1 = 5$	$8 \div 1 = 8$	$1 \div 1 = 1$	$3 \div 1 = 3$	1s
$24 \div 4 = 6$	$28 \div 4 = 7$	$20 \div 4 = 5$	$16 \div 4 = 4$	$32 \div 4 = 8$	4s
$30 \div 3 = 10$	$27 \div 3 = 9$	$12 \div 3 = 4$	$18 \div 3 = 6$	$24 \div 3 = 8$	3s
$36 \div 4 = 9$	$40 \div 4 = 10$	$32 \div 4 = 8$	$8 \div 4 = 2$	$28 \div 4 = 7$	4s
$24 \div 3 = 8$	$15 \div 3 = 5$	$30 \div 3 = 10$	$21 \div 3 = 7$	$18 \div 3 = 6$	3s
$3 \div 1 = 3$	$6 \div 1 = 6$	$9 \div 1 = 9$	$10 \div 1 = 10$	$4 \div 1 = 4$	1s
$10 \div 2 = 5$	$12 \div 2 = 6$	$14 \div 2 = 7$	$0 \div 2 = 0$	$6 \div 2 = 3$	2s
$12 \div 3 = 4$	$30 \div 3 = 10$	$18 \div 3 = 6$	$9 \div 3 = 3$	$24 \div 3 = 8$	3s
$40 \div 4 = 10$	$0 \div 4 = 0$	$12 \div 4 = 3$	$24 \div 4 = 6$	$16 \div 4 = 4$	4s
$0 \div 1 = 0$	$10 \div 1 = 10$	$7 \div 1 = 7$	$1 \div 1 = 1$	$5 \div 1 = 5$	1s
$6 \div 2 = 3$	$16 \div 2 = 8$	$18 \div 2 = 9$	$20 \div 2 = 10$	$12 \div 2 = 6$	2s
$15 \div 3 = 5$	$24 \div 3 = 8$	$30 \div 3 = 10$	$18 \div 3 = 6$	$0 \div 3 = 0$	3s
$20 \div 4 = 5$	$24 \div 4 = 6$	$36 \div 4 = 9$	$40 \div 4 = 10$	$4 \div 4 = 1$	4s

Practice Sheet 6s, page 18

1	3	2	5	9	4	10	6	8	7
3	6	8	1	5	9	7	2	10	4
0	5	2	9	1	8	4	7	3	10
5	2	9	1	8	4	10	3	6	7
4	1	7	2	10	8	3	0	5	9
10	8	9	3	6	7	5	4	2	1
8	4	1	9	2	10	6	7	3	5
6	7	3	4	10	9	8	5	1	2
2	9	5	6	7	3	1	8	4	3
7	10	4	10	3	2	9	1	7	6

Page 19

1	2	9	10	4
3	8	5	7	10
0	2	1	4	3
5	9	8	5	6
4	7	10	3	5
10	9	6	5	2
8	1	2	6	3
6	3	10	8	1
2	5	7	1	4
7	4	3	9	7
3	10	4	6	6
6	9	6	2	3
5	4	8	7	2
2	9	4	3	5
1	3	8	0	1
8	2	7	4	9
4	1	10	7	7
7	0	9	5	10
9	5	3	8	4
10	6	2	1	7

Practice Sheet 7s, page 20

1	2	5	3	10	9	6	7	8	4
3	6	9	1	5	10	4	7	2	8
4	7	3	9	1	5	10	2	6	0
7	3	5	1	9	6	8	10	2	4
2	5	6	7	10	1	4	8	9	3
0	6	2	10	5	1	9	3	7	4
8	4	7	6	3	10	5	2	9	1
5	1	4	2	4	7	3	1	10	2
6	7	8	4	2	3	2	5	1	5
9	8	1	5	4	2	1	6	3	10

Page 21

1	5	10	6	8
3	9	5	7	4
4	3	1	10	6
7	5	9	8	2
2	6	10	4	9
0	2	5	9	7
8	7	3	5	9
5	4	4	3	10
6	8	2	2	1
9	1	4	1	3
2	3	9	7	4
6	9	10	2	8
7	1	5	10	0
3	7	6	2	4
5	10	1	8	3
6	1	3	3	4
4	6	1	2	1
1	2	7	1	2
7	4	3	5	5
8	5	2	6	10

Review Sheet 5–7, page 22

2	9	6	7	7	8	8	5	7	9
3	1	8	7	6	6	7	9	1	8
10	2	2	7	8	5	3	6	10	7
10	1	1	3	6	9	4	10	9	1
10	9	2	10	4	5	10	3	9	3
2	1	8	3	8	5	4	3	2	8
10	1	2	7	7	5	5	3	2	1
8	1	10	3	6	5	6	7	2	3
3	7	2	9	4	6	6	5	8	1
7	4	6	3	8	5	5	7	4	9

Section Diagnostic Test, 5–7, page 23

$20 \div 5 = 4$	$25 \div 5 = 5$	$35 \div 5 = 7$	$45 \div 5 = 9$	$40 \div 5 = 8$	5s
$42 \div 6 = 7$	$54 \div 6 = 9$	$48 \div 6 = 8$	$30 \div 6 = 5$	$36 \div 6 = 6$	6s
$7 \div 7 = 1$	$21 \div 7 = 3$	$14 \div 7 = 2$	$28 \div 7 = 4$	$70 \div 7 = 10$	7s
$15 \div 5 = 3$	$5 \div 5 = 1$	$50 \div 5 = 10$	$10 \div 5 = 2$	$30 \div 5 = 6$	5s
$6 \div 6 = 1$	$60 \div 6 = 10$	$24 \div 6 = 4$	$18 \div 6 = 3$	$12 \div 6 = 2$	6s
$35 \div 7 = 5$	$49 \div 7 = 7$	$63 \div 7 = 9$	$56 \div 7 = 8$	$42 \div 7 = 6$	7s
$10 \div 5 = 2$	$20 \div 5 = 4$	$30 \div 5 = 6$	$40 \div 5 = 8$	$50 \div 5 = 10$	5s
$6 \div 6 = 1$	$18 \div 6 = 3$	$54 \div 6 = 9$	$30 \div 6 = 5$	$42 \div 6 = 7$	6s
$63 \div 7 = 9$	$49 \div 7 = 7$	$35 \div 7 = 5$	$21 \div 7 = 3$	$7 \div 7 = 1$	7s
$15 \div 5 = 3$	$25 \div 5 = 5$	$35 \div 5 = 7$	$45 \div 5 = 9$	$5 \div 5 = 1$	5s
$12 \div 6 = 2$	$60 \div 6 = 10$	$36 \div 6 = 6$	$24 \div 6 = 4$	$48 \div 6 = 8$	6s
$56 \div 7 = 8$	$42 \div 7 = 6$	$28 \div 7 = 4$	$14 \div 7 = 2$	$70 \div 7 = 10$	7s
$45 \div 5 = 9$	$35 \div 5 = 7$	$40 \div 5 = 8$	$30 \div 5 = 6$	$25 \div 5 = 5$	5s
$36 \div 6 = 6$	$12 \div 6 = 2$	$54 \div 6 = 9$	$60 \div 6 = 10$	$42 \div 6 = 7$	6s
$49 \div 7 = 7$	$21 \div 7 = 3$	$42 \div 7 = 6$	$35 \div 7 = 5$	$70 \div 7 = 10$	7s
$20 \div 5 = 4$	$50 \div 5 = 10$	$10 \div 5 = 2$	$5 \div 5 = 1$	$15 \div 5 = 3$	5s
$18 \div 6 = 3$	$48 \div 6 = 8$	$30 \div 6 = 5$	$24 \div 6 = 4$	$6 \div 6 = 1$	6s
$63 \div 7 = 9$	$28 \div 7 = 4$	$56 \div 7 = 8$	$7 \div 7 = 1$	$14 \div 7 = 2$	7s
$60 \div 6 = 10$	$48 \div 6 = 8$	$54 \div 6 = 9$	$42 \div 6 = 7$	$24 \div 6 = 4$	6s
$42 \div 7 = 6$	$49 \div 7 = 7$	$24 \div 7 = 3$	$56 \div 7 = 8$	$63 \div 7 = 9$	7s

Practice Sheet 8s, page 24

1	3	2	4	6	9	10	7	8	5
3	7	5	2	9	6	1	8	4	10
4	8	1	5	6	9	2	7	10	3
5	10	8	4	1	3	6	2	7	9
2	3	10	9	4	8	7	1	5	6
10	4	1	8	9	6	5	2	3	7
8	5	3	6	5	9	1	10	2	1
1	9	4	1	7	2	3	9	3	2
9	6	3	10	2	4	3	5	1	4
6	1	7	3	8	10	4	3	7	9

Page 25

1	2	6	10	8
3	5	9	1	4
4	1	6	2	10
5	8	1	0	7
2	10	4	7	5
10	1	9	5	3
8	3	5	1	2
1	4	7	3	3
9	3	2	0	1
6	7	8	4	7
7	9	3	9	10
3	4	9	7	5
8	2	6	8	10
7	5	9	7	3
10	4	3	2	9
3	9	8	1	6
4	8	6	2	7
5	6	9	10	1
9	1	2	9	2
6	10	4	5	4

Practice Sheet 9s, page 26

1	2	4	10	3	5	9	7	8	6
2	10	7	3	1	6	8	4	9	5
5	9	4	8	6	1	3	7	10	2
3	6	9	4	10	8	1	6	2	7
7	2	6	1	8	10	4	9	5	3
4	6	2	8	9	7	10	3	1	5
5	1	3	10	7	9	8	2	6	4
10	3	1	9	5	2	4	10	3	8
6	4	5	2	4	3	6	5	2	1
8	7	10	6	2	4	5	8	1	0

Page 27

1	4	3	9	8
2	7	1	8	9
5	9	6	3	10
3	4	10	1	2
7	2	8	4	5
4	3	9	10	1
5	1	7	8	6
10	5	0	2	3
8	6	2	5	8
2	2	5	7	0
10	5	6	4	1
9	10	1	7	8
5	8	8	6	4
2	1	10	9	5
6	4	7	3	3
1	8	9	2	7
3	0	2	10	2
4	10	3	5	5
7	6	4	1	6
6	10	4	6	2

Practice Sheet 10s, page 28

10	8	3	9	4	1	6	2	5	7
6	1	9	5	8	3	7	10	4	2
7	5	2	6	1	4	9	3	8	10
6	2	10	7	3	8	5	9	1	4
5	8	3	2	7	9	6	4	10	1
9	6	2	4	9	10	8	1	5	3
1	10	4	9	6	7	2	3	8	5
3	5	1	8	10	2	9	4	6	8
8	4	3	2	1	5	10	6	9	7
7	9	8	10	5	6	1	2	3	4

Page 29

1	2	7	8	9
3	10	6	1	5
4	3	10	5	6
8	9	4	2	7
9	4	1	8	2
10	1	3	6	4
7	3	5	10	9
2	4	9	5	8
5	6	7	4	2
6	2	4	9	10
1	5	10	3	4
10	2	4	9	8
9	8	7	2	1
2	1	6	10	3
8	10	5	3	7
6	5	9	2	9
5	8	1	4	6
9	6	3	1	10
7	9	8	3	1
6	3	7	8	5

Review Sheet 8–10, page 30

1	2	8	9	2
9	10	1	4	10
10	3	1	3	9
3	6	7	8	7
4	3	4	5	3
5	2	6	6	7
3	4	6	5	7
4	4	3	4	6
6	8	3	7	6
1	3	10	1	2
9	2	5	9	6
6	5	8	5	7
8	1	4	4	8
10	6	8	7	6
4	5	3	9	2
3	7	8	8	10
8	7	6	9	6
4	9	8	1	5
7	8	4	9	10
10	7	10	1	8

Section Diagnostic Test 8–10, page 31

9s	8s	10s	8s	9s	10s	8s	9s	10s	8s
9)9 = 1	8)64 = 8	10)100 = 10	8)48 = 6	9)90 = 10	10)20 = 2	8)56 = 7	9)36 = 4	10)100 = 10	8)48 = 6
9)81 = 9	8)48 = 6	10)20 = 2	8)64 = 8	9)81 = 9	10)40 = 4	8)8 = 1	9)72 = 8	10)70 = 7	8)24 = 3
9)63 = 7	8)32 = 4	10)50 = 5	8)56 = 7	9)72 = 8	10)60 = 6	8)80 = 10	9)81 = 9	10)50 = 5	8)56 = 7
9)45 = 5	8)16 = 2	10)60 = 6	8)40 = 5	9)63 = 7	10)80 = 8	8)24 = 3	9)63 = 7	10)20 = 2	8)32 = 4
9)27 = 3	8)80 = 10	10)80 = 8	8)64 = 8	9)81 = 9	10)100 = 10	8)40 = 5	9)54 = 6	10)40 = 4	8)64 = 8
9)18 = 2	8)8 = 1	10)90 = 9	8)72 = 9	9)45 = 5	10)10 = 1	8)48 = 6	9)63 = 7	10)80 = 8	8)72 = 9
9)36 = 4	8)24 = 3	10)70 = 7	8)24 = 3	9)36 = 4	10)30 = 3	8)72 = 9	9)36 = 4	10)60 = 6	8)56 = 7
9)54 = 6	8)40 = 5	10)40 = 4	8)32 = 4	9)27 = 3	10)50 = 5	8)64 = 8	9)54 = 6	10)30 = 3	8)72 = 9
9)72 = 8	8)56 = 7	10)30 = 3	8)48 = 6	9)18 = 2	10)70 = 7	8)32 = 4	9)81 = 9	10)90 = 9	8)40 = 5
9)90 = 10	8)72 = 9	10)10 = 1	8)32 = 4	9)9 = 1	10)90 = 9	8)16 = 2	9)45 = 5	10)70 = 7	8)64 = 8

Review Sheet 1–10, page 32

2	6	6	5	10
6	7	6	7	4
6	3	9	3	5
8	6	3	5	9
6	10	10	8	7
9	6	7	10	8
4	5	9	8	2
4	4	8	6	10
9	2	9	10	2
2	6	9	7	8
1	8	6	8	5
9	4	3	1	9
6	9	6	4	4
3	7	10	6	8
7	5	7	4	8
7	9	5	7	3
6	8	3	7	9
6	10	8	2	1
3	3	4	2	9
6	5	2	10	2

The *Final Assessment Test* is arranged like the Beginning Assessment Test, by fact diagonals.

Final Assessment Test 1–10, page 33

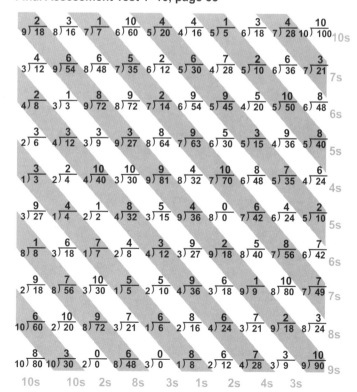

9)18 = 2	8)16 = 3	7)7 = 1	6)60 = 10	5)20 = 4	4)16 = 4	5)5 = 1	6)18 = 3	7)28 = 4	10)100 = 10	10s
3)12 = 4	9)54 = 6	8)48 = 6	7)35 = 5	6)12 = 2	5)30 = 6	4)28 = 7	5)10 = 2	6)36 = 6	7)21 = 3	7s
4)8 = 2	3)3 = 1	9)72 = 8	8)72 = 9	7)14 = 2	6)54 = 9	5)45 = 9	4)20 = 5	5)50 = 10	6)48 = 8	6s
2)6 = 3	4)12 = 3	3)9 = 3	9)27 = 3	8)64 = 8	7)63 = 9	6)30 = 5	5)15 = 3	4)36 = 9	5)40 = 8	5s
1)3 = 3	2)4 = 2	4)40 = 10	3)30 = 10	9)81 = 9	8)32 = 4	7)70 = 10	6)48 = 8	5)35 = 7	4)24 = 6	4s
3)27 = 9	1)4 = 4	2)2 = 1	4)32 = 8	3)15 = 5	9)36 = 4	8)0 = 0	7)42 = 6	6)24 = 4	5)10 = 2	5s
8)8 = 1	3)18 = 6	1)7 = 7	2)8 = 4	4)12 = 3	3)27 = 9	9)18 = 2	8)40 = 5	7)56 = 8	6)42 = 7	6s
2)18 = 9	8)56 = 7	3)30 = 10	1)5 = 5	2)10 = 5	4)36 = 9	3)18 = 6	9)9 = 1	8)80 = 10	7)49 = 7	7s
10)60 = 6	2)20 = 10	8)72 = 9	3)21 = 7	1)6 = 6	2)16 = 8	4)24 = 6	3)21 = 7	9)18 = 2	8)24 = 3	8s
10)80 = 8	10)30 = 3	2)0 = 0	8)48 = 6	3)0 = 0	1)8 = 8	2)12 = 6	4)28 = 7	3)9 = 3	9)90 = 10	9s

10s   10s   2s   8s   3s   1s   2s   4s   3s

Math Series

The Straight Forward Math Series

is systematic, first diagnosing skill levels, then *practice*, periodic *review*, and *testing*.

Blackline

GP-006 Addition
GP-012 Subtraction
GP-007 Multiplication
GP-013 Division
GP-039 Fractions
GP-083 Word Problems, Book 1
GP-042 Word Problems, Book 2

The Advanced Straight Forward Math Series

is a higher level system to diagnose, practice, review, and test skills.

Blackline

GP-015 Advanced Addition
GP-016 Advanced Subtraction
GP-017 Advanced Multiplication
GP-018 Advanced Division
GP-020 Advanced Decimals
GP-021 Advanced Fractions
GP-044 Mastery Tests
GP-025 Percent
GP-028 Pre-Algebra Book 1
GP-029 Pre-Algebra Book 2
GP-030 Pre-Geometry Book 1
GP-031 Pre-Geometry Book 2
GP-163 Pre-Algebra Companion
GP-168 Fractions Mastery

Upper Level Math Series

GP-104 Algebra, Book 1
GP-105 Algebra, Book 2
GP-167 Algebra Book 3
GP-045 Trigonometry
GP-054 Geometry
GP-053 Pre-Calculus
GP-064 Calculus AB, Vol. 1
GP-067 Calculus AB, Vol. 2

Math Pyramid Puzzles

Math Pyramid Puzzles
GP-162
5 two-sided puzzles

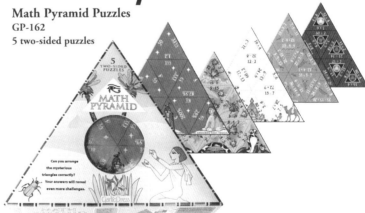

Assemble 5 two-sided puzzles each with different mathematical challenges. Solve the mathematical pyramid on the front side, turn the clear tray over to reveal a problem of logic: percents, decimals, fractions, exponents and factors.

Start building your pyramid at the bottom. The center piece is labeled and the picture may offer a clue.

Use your math skills to match sides with the same value.

You may find more than one match, but **all sides that touch** must match. When you are satisfied with your solution, close the tray.

Turn over and check the back. If the pieces are in order, you are correct!

Now, can you solve this logic puzzle?

ENGLISH SERIES

The **Straight Forward English** series is designed to measure, teach, review, and master specific English skills. All pages are reproducible and include answers to exercises and tests.

Capitalization & Punctuation
GP-032 • 40 pages
I and First Words; Proper Nouns; Ending Marks and Sentences; Commas; Apostrophes; Quotation Marks.

Nouns & Pronouns
GP-033 • 40 pages
Singular and Plural Nouns; Common and Proper Nouns; Concrete and Abstract Nouns; Collective Nouns; Possessive Pronouns; Pronouns and Contractions; Subject and Object Pronouns.

Verbs
GP-034 • 40 pages
Action Verbs; Linking Verbs; Verb Tense; Subject-Verb Agreement; Spelling Rules for Tense; Helping Verbs; Irregular Verbs; Past Participles.

Sentences
GP-041 • 40 pages
Sentences; Subject and Predicate; Sentence Structures.

Adjectives & Adverbs
GP-035 • 40 pages
Proper Adjectives; Articles; Demonstrative Adjectives; Comparative Adjectives; Special Adjectives: Good and Bad; -ly Adverbs; Comparative Adverbs; Good-Well and Bad-Badly.

Prepositions, Conjunctions and Interjections
GP-043 • 40 pages
Recognizing Prepositions; Object of the Preposition; Prepositional Phrases; Prepositional Phrases as Adjectives and Adverbs; Faulty Reference; Coordinating, Correlative and Subordinate Conjunctions.

ADVANCED ENGLISH SERIES

Get It Right!
GP-148 • 144 pages
Organized into four sections, **Get It Right!** is designed to teach writing skills commonly addressed in the standardized testing in the early grades: Spelling, Mechanics, Usage, and Proofreading. Overall the book includes 100 lessons, plus reviews and skill checks.

All-In-One English
GP-107 • 112 pages
The **All-In-One** is a master book to the Straight Forward English Series.
Under one cover it has included the important English skills of capitalization, punctuation, and all eight parts of speech. Each selection of the All-In-One explains and models a skill and then provides focused practice, periodic review, and testing to help measure acquired skills. Progress through all skills is thorough and complete.

Grammar Rules!
GP-102 • 250 pages
Grammar Rules! is a straightforward approach to basic English grammar and English writing skills. Forty units each composed of four lessons for a total of 160 lessons, plus review, skill checks, and answers. Units build skills with Parts of Speech, Mechanics, Diagramming, and Proofreading. Solid grammar and writing skills are explained, modeled, practiced, reviewed, and tested.

Clauses & Phrases
GP-055 • 80 pages
Adverb, Adjective and Noun Clauses; Gerund, Participial and Infinitive Verbals; Gerund, Participial, Infinitive, Prepositional and Appositive Phrases.

Mechanics
GP-056 • 80 pages
Abbreviations; Apostrophes; Capitalization; Italics; Quotation Marks; Numbers; Commas; Semicolons; Colons; Hyphens; Parentheses; Dashes; Brackets; Ellipses; Slashes.

Grammar & Diagramming Sentences
GP-075 • 110 pages
The Basics; Diagramming Rules and Patterns; Nouns and Pronouns; Verbs; Modifiers; Prepositions, Conjunctions, and Special Items; Clauses and Compound-Complex Sentences.

Troublesome Grammar
GP-019 • 120 pages •
Agreement; Regular and Irregular Verbs; Modifiers; Prepositions and Case, Possessives and Contractions; Plurals; Active and Passive Voice;